CONTENTS

Boldface words appear in the glossary.

A Cheerful Bird

Who's that bird with the cheerful song?

It's an American robin. Let's sing along!

This sign of spring is a welcome surprise.

They start each day with the sunrise.

You can sometimes find them in a nest.

Each has a brightly colored chest!

ANIMAL RHYME TIME!

RHYME TIME WITH ROBINS!

BY LEIGH MCCLURE

Gareth Stevens PUBLISHING

Please visit our website, www.garethstevens.com. For a free color catalog of all our high-quality books, call toll free 1-800-542-2595 or fax 1-877-542-2596.

Library of Congress Cataloging-in-Publication Data

Names: McClure, Leigh, author.
Title: Rhyme time with robins! / Leigh McClure.
Description: Buffalo, NY : Gareth Stevens, [2024] | Series: Animal rhyme time! | Includes index.
Identifiers: LCCN 2022051548 (print) | LCCN 2022051549 (ebook) | ISBN 9781538285367 (library binding) | ISBN 9781538285350 (paperback) | ISBN 9781538285374 (ebook)
Subjects: LCSH: Robins–Juvenile literature. | Rhyme–Juvenile literature.
Classification: LCC QL696.P288 M39 2024 (print) | LCC QL696.P288 (ebook) | DDC 598.8/42–dc23/eng/20221117
LC record available at https://lccn.loc.gov/2022051548
LC ebook record available at https://lccn.loc.gov/2022051549

Published in 2024 by
Gareth Stevens Publishing
2544 Clinton Street
Buffalo, NY 14224

Designer: Claire Wrazin
Editor: Caitie McAneney

Photo credits: Cover Dominique Bradette/Shutterstock.com; p. 5 Kit Leong/Shutterstock.com; p. 7 Jukka Jantunen/Shutterstock.com; p. 9 (main) Gerald A. Deboer/Shutterstock.com; p. 9 (inset) Nick Vorobey/Shutterstock.com; p. 11 Ayman Haykal/Shutterstock.com; p. 13 Chris Higgins Photography/Shutterstock.com; p. 15 RobDun/Shutterstock.com; p. 17 Evgeny Haritonov/Shutterstock.com; p. 19 Anne Kitzman/Shutterstock.com; p. 21 Karel Bock/Shutterstock.com.

Printed in the United States of America

CPSIA compliance information: Batch #CS24GS: For further information contact Gareth Stevens, New York, New York at 1-800-542-2595.

What's a Robin?

American robins are in the thrush family.
They hop on the ground or sit in a tree.
Each may grow to 10 inches long.
This bird has a special springtime song.
Its chest has orange or red feathers.
It hunts for worms in warm, wet weather.

Where on Earth?

Two kinds of robins live on Earth.

They're named for their place of birth.

European robins are small and round.

Europe or Africa are where they're found.

As for American robins, can you guess?

They're in Canada, Mexico, and the U.S.

AMERICAN ROBIN
EUROPEAN ROBIN

Early Bird Gets the Worm

Go in your backyard and look around.

A robin may be poking at the ground!

To robins, a rainy day is a great deal

because worms are their favorite meal!

Using **beaks**, they get worms by the tail!

They'll also eat **insects**, fruit, and snails.

Baby Robins

Robins lay eggs like all birds do.

Their eggs are small and light blue.

A bird's eggs are very dear.

Parents fight predators that come near.

Eggs rest in a nest of mud and wood.

Soon, babies will break out for good!

Growing Up

Three to five eggs are in each **brood**.

Baby birds **hatch**, then cry out for food!

Hatchlings have no feathers or sight.

Nestlings are not yet ready for flight.

Mama brings food to get them by.

Within a few weeks, they're ready to fly!

NESTLINGS

Ready to Fly

Fledglings can hop, but they're still trying
to get the hang of really flying!
They leave the nest to find a meal,
but stay close by 'til they can fly for real.
If you see a fledgling, leave it be!
It learns by hopping around, you see.

Home Tweet Home

Robins make their nests high in a tree.

This keeps babies as safe as can be.

Squirrels and big birds are a **threat**,

mostly to babies that can't fly just yet.

Robins **roost** in groups in cold weather.

They'll ride out the long winter together!

A Sign of Spring

Some robins fly south when things freeze.
Others spend winter roosting in trees.
They look around for berries to eat.
In winter, it's hard to find a treat.
Often the first to lay eggs each year,
a robin's song is sure to spread cheer!

GLOSSARY

beak: A part of the mouth that sticks out on some animals and is used to tear food.

brood: A group of young animals that were all born at the same time.

fledgling: A young bird that only recently gained the ability to fly.

hatch: To break open or come out of.

insect: A small, often winged, animal with six legs and three main body parts.

roost: To settle down for rest or sleep.

threat: Something likely to cause harm.

FOR MORE INFORMATION

BOOKS

London, Martha. *Robins*. Lake Elmo: MN: Focus Readers, 2021.

Nelson, Kristen Rajczak. *Robins*. Buffalo, NY: Gareth Stevens Publishing, 2023.

WEBSITES

American Robin
kids.nationalgeographic.com/animals/birds/facts/american-robin
Explore more about the American robin with National Geographic Kids.

American Robin Facts for Kids
naturemappingfoundation.org/natmap/facts/american_robin_k6.html
Find more facts about robins, including how to identify them.

Publisher's note to educators and parents: Our editors have carefully reviewed these websites to ensure that they are suitable for students. Many websites change frequently, however, and we cannot guarantee that a site's future contents will continue to meet our high standards of quality and educational value. Be advised that students should be closely supervised whenever they access the internet.

INDEX